Rare Earth Ramble
Collected Poems 2022-2023

by Rick Bernstein

Rare Earth Ramble: Collected Poems 2022-2023

Introduction

Rare Earth Ramble presents a selection of poems written in 2022-2023, including poems about the chemical elements. This book is the fourth in a series. The first is *Greasy Not Green,* the second *Chaos at the Café* with artist Margaret Bernstein, and the third *Poems of the Periodic Table.* *Rare Earth Ramble* includes more poems about elements, as well as poems about people, and a few poems about science and other amusing topics. The reader may find a favorite or two.

    Rick Bernstein, Las Cruces, New Mexico, July 2023

Dedicated to Margaret, as always.

	page
Table of Contents	
Title page	1
Introduction and Dedication	2
Table of Contents	3
Section 1: Poems Elemental	
Introduction to the Periodic Table	6
Four Ancient Elements	7
Some Elements: Niobium, Technetium, Indium, Tellurium, Hafnium, Iridium, Thallium	9
Atomic symbols Nb,Tc,In,Te,Hf,Ir,Tl	
Atomic numbers 41,43,49,52,72,77,81	
Rare Earth Elements:	17
Scandium, Yttrium	17
Atomic symbols Sc, Y numbers 21, 39	
The Lanthanides	19
Elements: Lanthanum, Cerium, Praseodymium, Neodymium, Promethium, Samarium, Europium, Gadolinium, Terbium, Dysprosium, Holmium, Erbium, Thulium, Ytterbium, Lutetium	20
Atomic symbols La,Ce,Pr,Nd,Pm,Sm,Eu, Gd,Tb,Dy,Ho,Er,Tm,Yb,Lu numbers 57-71	
Title Poem: Rare Earth Ramble	34
Heavier Elements: The Actinides	39
Elements Actinium, Neptunium, Americium, Berkelium, Californium	40
Atomic symbols Ac, Np, Am, Bk, Cf	
Atomic numbers 89, 93, 95, 97, 98	
More Actinides	45

Table of Contents, continued   page

   Radioactive Elements of the Periodic Table
      Named for Human Beings   47
Poems of Radioactive Elements Named for People   49
    Curium, Einsteinium, Fermium, Mendelevium,
    Nobelium, Lawrencium, Rutherfordium,
    Seaborgium, Bohrium, Meitnerium, Roentgenium,
    Copernicium, Flerovium, Oganesson
      Atomic numbers 96, 99, 100, 101, 102, 103,
        104, 106, 107, 109, 111, 112, 114, 118
Radioactive Elements Named for Places   62
U-238 Decay Chain   64
Up and Down the Decay Chain   65
Decay Chain Elements: Uranium (Part 1),   67
    Thorium, Protactinium, Radium, Radon,
    Astatine, Polonium, Bismuth, Lead
    Atomic symbols U,Th,Pa,Ra,Rn,At,Po,Bi,Pb
    Atomic numbers 92,90,91,88,86,85,84,83,82
Atomic Bomb Elements   78
    Uranium (Part 2), Plutonium
        Atomic symbols, U, Pu   numbers 92, 94
Atomic Bombs   80
Fission and Fusion   81

Table of Contents, continued   page

Section 2: People
    Bad Luck Blues   83
    Second   85
    Great Men of Cigars   87
    Double Oboe Concerto   89
    How Many Was That Again?   93
    Aaron and Ara   95
    Freshmen Heavyweight Crew   97
    80 and More   100
    Co … py   101
    Lost Boys   102

Section 3: Miscellany, Fun and Serious
    Purpose of Life: Science and the Chicken   104
    Bird Song   106
    On the Air   108
    Four Rhymes   109
    Exercise in Style   111
    Reimagining the Alphabet   112
    Nantucket Limericks   114
    More Limericks   116
    Xmas Gift Poems   118

Front cover art and other cover illustrations courtesy of artist Margaret Bernstein.

Section 1: Poems Elemental

Introduction to the Periodic Table

All 118 known chemical elements are arranged in the Periodic Table in rows and columns showing their relatedness. The table is usually attributed to Dmitri Mendeleev, because his 19th Century arrangement of the known 60 or so elements at the time foreshadowed and predicted the chemical and physical properties of the later discovered remaining elements. In *Rare Earth Ramble* the first section contains poems about chemical elements. The previous book, *Poems of the Periodic Table*, covered mainly the lighter better-known more common elements plus some heavier radioactive elements. The present book deals mainly with the heavier, rarer elements, especially the rare earth elements and radioactive synthetic elements.

Four Ancient Elements

Our ancestors counted four:
Water, air, fire, and earth.
We have a different view today.
We count 118 at present, though
Only about 86 occur naturally,
The others synthetic, just
Tiny amounts in physics labs.

How well do the ancient four
Match today's 118, or match
More realistically, today's 86?

Water is 100% H2O, just two elements,
Hydrogen and oxygen,
Common, essential for life.
Good choice for a primitive element.
Air is more than 98% just two:
Nitrogen and oxygen.
Another good choice for
A basic element, essential for life.

No problem with fire: a candle flame
Or gas flame is just incandescent
Carbon particles, symbol C.
Soot, charcoal from fire
Are almost pure carbon C.

Coal and coke to produce fire
Likewise more than 95% carbon.

Seems like the ancients were almost right.
These three classical elements
Almost purely determine
Four of the six most essential elements
For life on Earth: H, O, N, C, P, and S,
Missing only phosphorus and sulfur.
These six modern elements make up
Virtually all biological molecules,
Account for more than 90% of biomass.

What about the fourth
Ancient element: earth?
That's where we get P, S, and most of
Remaining 80 or so of today's
Naturally occurring elements
That comprise our universe,
Our planet and all life on it.

The ancients were not far wrong,
Prescient in a simpler time.
The classical four elements lasted
Over 2000 years, pre-Socrates
To the Renaissance.
A different kind of science
From what we know today.

Some Elements: Niobium, Technetium, Indium, Tellurium, Hafnium, Iridium, Thallium
Atomic symbols Nb, Tc, In, Te, Hf, Ir, Tl
Atomic numbers 41, 43, 49, 52, 72, 77, 81

Niobium

Niobe daughter of Tantalus
Metal niobium very similar to tantalum
Niobium atomic number 41, symbol Nb
73-tantalum is larger by an entire
32-electron shell, lies under
Nb in same column of the table.
Niobium found in ores like columbite,
Explains original name of Nb, columbium,
Name still used in U.S. metallurgy.
Nb good alloy, hardens steel,
Temp-stable for jet engines,
Substitutes for allergenic nickel in jewelry.
Just imagine niobium jewelry
Decorating fair Queen Niobe,
Turned to niobium stone,
Weeping niobium tears
For her slain mythical children.

Technetium

Where is technetium?
Dmitri Mendeleev said it was there,
But nobody could find it.
Many false claims of discovery
To fill the 43-hole in the table.
Dmitri described its physical
And chemical properties in 1869.
Atomic number 43, symbol Tc
Chemists frustrated not to find technetium.
Tc very rare because of isotope decay,
Finally found in 1937 in cyclotron test,
Tc thus the first synthetic element.
Named for Greek *technetos*: artificial.
Where is Tc today?
In supernovas: inaccessible to humans.
On Earth, remnants of radioactive decay
From uranium and thorium:
Too rare, too little to find in nature.
On a small piece of molybdenum foil
From UCBerkeley cyclotron, where
Tc now long since decayed.
Tc isolated from spent nuclear fuel rods,
Today's only source.
Lowest atomic number of all
Purely radioactive elements
Having no non-radioactive isotopes.

Promethium similarly, at atomic number 61:
No non-radioactive isotopes.
All other purely radioactive elements
Lie beyond bismuth, number 83 in the table.
Tc lightest element with all isotopes radioactive.
Isotope Tc-99 weak beta emitter,
Good standard for calibration.
Tc radioactive tracer for nuclear medicine.
Tc minor alloy in steel, prevents rust
For steel used away from living beings.
Tc chemistry like manganese, several oxidation states.
Compounds like Tc+4 salts $TcCl_4$, and sulfide TcS_2,
Pertechnetate TcO_4 and organometallics.
Lots of chemistry known for an element
Present in supernovas but
Vanished from the primordial Earth.

Indium

Indium a soft silvery metal,
Similar to tin in properties.
Atomic number 49, symbol In
Very low melting point, like gallium and thallium,
Ga, Th each a homolog above or below,
In same column of the Periodic Table.
Indium identified by very bright
Violet spectral emission line.
Named from *indicum*, violet in Latin.
Indium almost melts in your hand.
Appropriate for low-melting alloys,
Especially solders, fire sprinklers.
Indium coats glass, useful in semiconductors.
Indium atom's fifth shell with
Three outer electrons, called 5s2,5p1.
When all three are donated in reactions,
Indium atom is trivalent In+3.
When only the 5p electron is donated,
The In atom reacts with +1 valence.
Explains much of indium chemistry.
No biological role, low toxicity of
Indium and In compounds.
Unusual distribution of In isotopes:
Radioactive isotope In-113 more abundant
Than the only stable isotope In-115.
True only for indium, rhenium, tellurium.
Among all 118 elements in the table.

Tellurium

Lost to Earth as a gas
Gone to outer space
In the hot millennia of Earth's
Formation as a planet.
Tellurium silvery white metal
Atomic number 52, symbol Te
Te as rare in Earth's crust as platinum,
Though common elsewhere in universe.
Reacts easily with hydrogen
To form volatile Te hydride,
Lost from the crust into the atmosphere.
Tellurium chemically similar to
Sulfur and selenium as a group,
Oxidation state valences -2, +2, +4, +6,
In chalcogen chemical family including
Oxygen and polonium, all lacking
Two electrons to fill outermost shell.
Te in semiconductors and solar panels.
Sulfur-containing amino acids
Cysteine and methionine can substitute
Te for S (also selenium Se for S).
Tellurium mildly toxic to humans.
Diagnose toxicity in people:
Tastes metallic, breath smells like garlic.

Hafnium

Shiny silvery gray metal
Atomic number 72, symbol Hf
Hf chemically similar to zirconium,
Both with +4 valence.
Hf found with zirconium in ores,
Difficult to separate and distinguish.
Main difference is neutron cross-section:
Hf greatly absorbs neutrons,
Good for nuclear control rods,
Prevents runaway reactors.
Zirconium is neutron-transparent,
Good for cladding on nuclear fuel rods,
Makes neutrons very available to react.
Hf named after Latin for Copenhagen,
Discovered there in 1923 though predicted
Much earlier in 1869 by Mendeleev.
Hf hard but ductile, wires for
Electronic circuits, filaments.
Forms superalloys with titanium,
Tantalum, niobium, or tungsten.
Strong, heat-resistant, for rocket engines.
Note the "Hafnium controversy":
Can an Hf isotope make a nuclear bomb?
Probably not, but don't ask the "Hafnium,"
Group of expert cyber hackers carrying out
Espionage for the Chinese government.

Iridium

Iridium a silvery hard dense metal,
Denser and more unreactive than gold.
Atomic number 77, symbol Ir
More highly resistant to corrosion
Than gold: Ir resists aqua regia.
Very high melting point, difficult to form.
Platinum-iridium alloys very stable,
Used to create the kilogram standard.
Ir also makes nibs for fountain pens,
High-temp crucibles for smelting.
1957, Rudolf Mössbauer showed emission and
Re-absorption of gamma rays by
Iridium isotope Ir-157. For revolutionary
"Mössbauer effect" he won
Nobel Prize in Physics in 1961 at age 32.
Iridium also the name for
Group of communication satellites,
Though not made from iridium metal.

Thallium

The "poisoner's poison,"
Thallium a favorite of Agatha Christie
In her drawing room mysteries.
Atomic number 81, symbol Tl.
A gray-white metal,
But found in nature only chemically bound
With other elements as ores.
Thallium mimics potassium chemically.
Salts of thallium very toxic,
Used as rodenticides and insecticides,
Poisonous to rats and people.
Get rid of the rat in your life.
Called the "inheritance powder"
Tl salts water-soluble, tasteless, odorless.
Question: accident or deliberate criminal act?
Hard to tell.

Rare Earth Elements:
    Scandium, Yttrium
        Atomic symbols Sc, Y   numbers 21, 39
    The Lanthanides
    Poems of the Lanthanide Elements:
        Lanthanum, Cerium, Praseodymium,
        Neodymium, Promethium, Samarium, Europium,
        Gadolinium, Terbium, Holmium, Dysprosium,
        Erbium, Thulium, Ytterbium, Lutetium
            Atomic symbols La, Ce, Pr, Nd, Pm, Sm,
                Eu, Gd, Tb, Dy, Ho, Er, Tm, Yb, Lu
Atomic numbers 57-71

Scandium

Discovered in Scandinavian ores in 1879,
Predicted by Mendeleev in 1869.
Atomic number 21, symbol Sc
Sc next to calcium along the row,
The row of transition metals,
Among most important metals on earth:
Ti, V, Cr, Mn, Fe, Co, Ni, Cu, and Zn at 30.
All common minerals and very useful metals,
Though scandium has limited uses.
Rare earth with yttrium and lanthanides.
No role for scandium in biology,

Yet absorbed from soil and water
Into plants and fish, and then to animals.
Fortunately non-toxic.
Scandium is lightweight,
Strong alloys with aluminum for
Fishing rods and golf clubs,
Baseball bats and bike frames.
Play sports, have fun with scandium.

Yttrium

A rare earth mineral,
Atomic number 39, symbol Y (only)
The first in row of transition series metals.
Often found in nature with
Lanthanides and other rare earths.
Y precedes zirconium Zr in table.
Y used in electronics, phosphors,
Red color in early CRT color TVs,
Now in today's LEDs.
Y alloys for strength with
Magnesium and aluminum.
Yttrium aluminum garnet YAG, synthetic
Diamond, hard and sparkling, brilliant.
As rare earth, Y short in supply,
But not as rare as real diamonds.

The Lanthanides

A series of 15 metal elements,
Atomic numbers 57-71,
Starting with lanthanum, symbol La,
Then cerium Ce, praseodymium Pr,
Neodymium Nd, promethium Pm,
Samarium Sm, europium Eu,
Gadolinium Gd, terbium Tb,
Dysprosium Dy, holmium Ho,
Erbium Er, thulium Tm,
Ytterbium Yt, finally lutetium Lu.
Termed rare earth elements,
Similar chemistries, trivalent from
Behavior of common 4f orbital electrons.
Also chemistry similar to scandium Sc
And yttrium Y, numbers 21 and 39,
Both termed rare earths with trivalency.
Many practical uses in technology,
Especially lasers, magnets, catalysts,
But no biological roles, though possibly
Anti-cancer drugs based on La and Ce.
Supply of rare earths a critical issue
For national security, they say.

15 Lanthanide Elements:
   Lanthanum, Cerium, Praseodymium,
   Neodymium, Promethium, Samarium, Europium,
   Gadolinium, Terbium, Holmium, Dysprosium,
   Erbium, Thulium, Ytterbium, Lutetium

Lanthanum

The first of 15 elements of the
Lanthanide series of rare earths,
Series named for lanthanum
As the first and prototype.
Atomic number 57, symbol La.
Relatively abundant, found with
Other rare earths especially cerium.
Ductile silvery metal, softest of
The lanthanides, which get harder
Marching through the series.
Many uses for lanthanum in
Electronics, arc lights, electrodes,
Catalysts, similar for other rare earths.
Coleman gas lamp mantles shine
Brightly with lanthanum.
No biological role in humans, no toxicity.
Some methane-producing bacteria
Depend on essential La for enzymes,
But other lanthanides can substitute.

Cerium

Cerium, a rare earth metal element
Second in the lanthanide series
Following lanthanum itself.
Most common lanthanide in Earth's crust,
Even five times more common than lead.
Atomic number 58, symbol Ce
Ce discovered in 1803, first of the 15.
Very unusual properties,
Four allotropes and four stable isotopes,
Metal allotropes have different crystal structures.
Isotopes theoretically radioactive
But so long-lived effectively stable.
Ce+4 oxidation state unusual among +3 lanthanides.
The oxide ceria approximates CeO_2
Easily extracted from ores in water.
Ce prime ingredient of famous
Belousov-Zhabotinskyy reaction:
Ce oscillates between +3 and +4 oxidation states,
Creating spatial and temporal arrangements
Starting from homogeneous aqueous solution,
Spontaneously forming shapes and
Patterns in liquid.
A model for biological differentiation,
Like fertilized egg, a single cell,
Becoming complex multicellular tissues,
In organs, spatially arranged.
B-Z reaction has to be seen to be believed.

Praseodymium

Intense yellow colored glass,
Filters bright yellow and infrared,
Welder's glasses with neodymium
In "didymium" combination that
Occurs naturally in ores.
Rare earth lanthanide praseodymium,
Atomic number 59, symbol Pr
Soft silvery metal, malleable and ductile.
Pr useful magnetic properties from
Unpaired 4f electron,
Like some other lanthanides.
Lasers, fiber optics, photonics, phosphors
Of several colors.
Powerful small magnets with
Neodymium and other lanthanides.
No role in biology, except
Some methane-producing bacteria.

Neodymium

Lanthanide, rare earth, silvery metal
Like 14 other lanthanides, all similar
Neodymium atomic number 60, symbol Nd.
Hard to distinguish chemically from
Number 59 praseodymium Pr.
Together Nd and Pr were called "didymium,"
Thought to be a unique element until 1940s.
Didymium still used in US for special glasses.
Nd many uses: lasers, colored art glass.
Alloyed with iron and boron to make
Small, very powerful permanent magnets for
Earphones, watches, pacemakers, cell phones,
Small powerful motors, good for space craft,
Very small magnets, called Zen magnets,
Neoballs, and buckyballs for children's toys.
If a child swallows more than one,
Causes gut tissues to pinch together,
Become necrotic, requiring emergency surgery.
More than 1700 emergencies in US children.
Buckyballs have been recalled,
Now prohibited in toys.

Promethium

Hail Prometheus, Bringer of Fire!
Promethium named for Greek Titan,
Stole fire from Mount Olympus,
Brought fire to humans.
Atomic number 61, symbol Pm
Fission product of uranium,
All 38 Pm isotopes radioactive.
Rarest rare earth metal, lanthanide series,
Only about 500 grams on Earth.
Promethium and technetium, only two
Radioactive elements followed by
Non-radioactives in the Periodic Table.
Pm chemistry strictly based on +3 reactivity.
Very few practical uses, mostly research,
Pm-147 only practical isotope of Pm,
Makes miniature batteries
For pacemakers, spacecraft.
It carries fire like Prometheus.

Samarium

Rare earth lanthanide samarium.
Found worldwide in ores cerite, gadolinite,
Monazite, bastnäsite, and samarskite.
Atomic number 62, symbol Sm
Sm magnets small and powerful
Especially when alloyed with cobalt,
Perform at high temperature,
Also used in guided missiles.
Trivalent chemistry like all lanthanides,
Many useful Sm compounds.
Samarium radioactive isotopes:
Sm-153 in multiple cancer treatments,
Sm-149 in nuclear reactor control.
Samarium named for mineral samarskite
From mine in Ural Mountains in 1840,
Ore's name from Vassili Samarsky,
Russian mining engineer supervisor.
Sm first element named,
Though indirectly, for a person.

Europium

Most reactive of lanthanide series,
Silvery white metal, europium
Oxidizes in air, turns brown.
Atomic number 63, symbol Eu
Named for the continent.
Soft metal, easily machined,
Cuts with a knife.
Eu metal absorbs neutrons,
Make effective control rods for
Nuclear reactors, prevent runaways.
Eu trivalent +3 chemistry
But also divalent +2 salts, giving various
Useful colored compounds,
Halides, oxides, sulfides,
Even organoEu compounds,
Some for lasers, microelectronics,
Photonics and bright lights
Of all colors,
Like other rare earths.

Gadolinium

Gray silvery metal,
A rare earth element in the
Lanthanide series: numbers 57-71,
Like neodymium and promethium.
Symbol Gd, atomic number 64.
Gd is a green phosphor.
You may have it nearby
If you have an old color TV,
Cathode rays hit Gd pixels
To produce green colored light.
Gd used also as contrast agent
For MRI, to reduce water molecules
In living tissue for better imaging.
Many Gd isotopes, especially Gd-148
For radioisotope thermoelectric generators,
But Gd-148 so far uneconomical.
Many specialized uses for Gd: rust-resistant,
Alloys, paramagnetic, neutron targeting,
Nuclear shielding, even makes garnets.
Gadolinium is Jack of All Trades.

Terbium

Name from Swedish village
Ytterby, near gadolinite mine,
Ore source of several rare earths.
Terbium atomic number 65, symbol Tb
Trivalent chemistry +3 salts,
Some are phosphorescent lime-green,
Contrast with erbium rosy pink.
Phosphors for smartphone screens,
Along with dysprosium red and yellow.
They light up your cell phone.

Dysprosium

Soft bright lustrous metal
Lanthanide, rare earth
Atomic number 66, symbol Dy
Captures neutrons well, used
In control rods for nuclear reactors.
Easily machined into components,
Good for high-strength magnets,
Like holmium, another magnetic rare earth.
Dy salts like halides for
Lasers, phosphors, and bright lamps.
Dysprosium is difficult to isolate
And purify from ores and salts.
Dy name in Greek means "hard to get."

Holmium

Rare earth lanthanide
Atomic number 67, symbol Ho
Ho metal, neutron absorber, moderates
Nuclear reactions in power plants.
Ho alloys make strong magnets.
Ho solid-state lasers in medicine.
Ho found in several ores with other
Rare earth lanthanide elements.
Like other rare earths, Ho salts colored,
Yellow decorates art glass, and
Colors cubic zirconium pink.
Ho oxide called holmia as extracted from ore,
Similarly ceria, terbia, erbia, ytterbia.
Holmia highly paramagnetic, for magnets.
Holmium named for Latin name of Stockholm,
Not named for Sherlock.

Erbium

Atomic number 68, symbol Er
Er compounds 3+ chemistry,
Colored rosy pink, for ceramics and
Especially glass, not just for art but
Technology of fiber optics and lasers.
Pink color has ideal emission wavelengths.
Er laser 2940 nm wavelength emits light
Good for lasers in dentistry, also
Minimally invasive surgery, skin treatments.
Erbium and terbium, confusingly similar
In many ways, not just in names.
As rare earth lanthanides both
Names from Ytterby, Swedish village.
Ores mined around 1860 led to
Oxides erbia and terbia, which were
Mislabeled and switched identities.
Erbia was terbia, terbia was erbia.
Not cleared up until almost 20 years later.

Thulium

Soft silvery gray metal
Thulium also rare earth lanthanide.
Atomic number 69, symbol Tm
Tm first isolated in 1879 from ore
As oxide thulia with erbia and holmia,
All rare earth oxides, numbers 67-69.
Like other rare earths, thulium
In magnets, lasers, colored glass,
Ferrite coils that mix Tm with ferric oxide,
Used in microwave ovens and electronic circuits.
Tm like scandium used for arc lighting.
Tm in Euro banknotes to foil counterfeiting.
Thulium named for Thule, as Greek
Place name for Iceland.
Well-known U.S. Thule air force base
Located in Greenland
North of the Arctic Circle,
Now called Pituffik air base, not Thule.

Ytterbium

Iconic lanthanide, named for Ytterby,
Village in Sweden with gadolinite mine
Rich with multiple rare earth ores.
Atomic number 70, symbol Yb.
Ytterbium contrasts with yttrium:
Much heavier metal, salts toxic,
Salts possibly carcinogenic,
Yb somewhat radioactive
From naturally occurring isotopes.
Used as gamma ray sources in research.
Yb dopant to make stainless steel.
Yb also makes very reliable atomic clocks.
Village Ytterby also source of names
For yttrium, terbium, erbium,
All of them rare earth elements.
Four names from one name,
What's in a name?
Village Ytterby now famous in certain circles.

Lutetium

Densest, hardest lanthanide metal,
Highest melting point of lanthanides.
Atomic number 71, symbol Lu
Finishes the rare earths, begins next
The 6th-transition series of metals,
Like nearby hafnium and tantalum.
Strictly +3 reactivity, limited chemistry,
In that way no different from
Other 14 lanthanides.
Not many practical uses for Lu,
Though stable Lu a catalyst to
Crack petroleum for refining.
Radioactive Lu-176 can date meteorites.
Isotope Lu-177 good for imaging neural tumors.
Celebrate lutetium as dense, hard, and modest.

Title poem:
Rare Earth Ramble

Rare earth metals are not so rare.
You find them almost everywhere,
You will, if you labor in Chin-
A, working in a rare earth mine.
Most of them are lanthanides,
A row of fifteen metals, slides
Across the periodic chart:
Pay attention and be smart:
Start at 57, stable
At the midpoint of the table.
Begin with lanthanum, and run
To 70-ytterbium.
That's fourteen lanthanides
Sometimes lutetium besides
At 71. Then add two more
To total seventeen full score
Light 21-scandium fine
And yttrium at 39.
Both metals but not lanthanide,
Early in the table reside.
Rare earth metals, most in sequence,
Sum total seventeen, makes sense.
In this way the list thus settles.
All rare earths are useful metals
Lustrous silvery but not gems

Their wider application stems
From electrons jumping in their shells,
Allows semi-conductor wells,
Emitting light of rainbow hues,
Colors ceramics, glasses: blues,
Greens, purples, pink, lasers and light,
Electronics, batt'ries, all right.
Rare earths order el'mental
Here, by atomic numbers, tell:

Scandium lightweight alloys
With aluminum for toys,
Bike frames, race cars, and fishing rods
For catching bass and carp and cods.

Yttrium, red colors please
For LEDs and old TVs.

Lanthanum first lanthanide
Softest metal, note with pride
For catalysts, electrodes fine.
Lanthanum gas mantles shine.

Cerium with clever spatial
Colored patterns, looks palatial,
Yields crystals of four allotropes.
Also four stable isotopes.

Praseodymium glasses wear,
Protects and shields the eyes from glare.

Neodymium alloys with iron,
Small strong magnets, just add boron.

Promethium rarest rare earth,
Radioactive by god's mirth:
All 38 isotopes are hot.
What hath Prometheus thus wrought?

Samarium for carbon arcs
And glass to block bright welding sparks.

Europium with its green glow
Makes light from other metals show.

Gadolinium green light plies
In pixels, also MRIs.
Thermoelectric power lies
Still in future distant bright skies.

Terbium with phosphors green
Illuminate your TV screen.

Dysprosium neutron captures
Controls nuclear reactors.

Holmium for magnet, laser,
Corrodes in water, burns in air.

Erbium ions make pink glass
And radiate in laser blasts.
Good for dental surg'cal tricks
And to send through fiber optics.

Thulium lasers solid state
And portable x-rays create.

Ytterbium traces in steel
Help make a stainless alloy real.

Lutetium, last of lanthanides,
First transition 6th besides.
Still a rare earth on the lists,
Mainly for alloys, catalysts.

Ytterby, village in Sweden,
Four rare earth names have been given:
Ytterbium and yttrium,
Then terbium and erbium,
Nearby Ytterby you will find.
Mineral gadolinite mine
Whose ore contains ytterbium
Locked up with others in a scrum.

The seventeen are much the same
From intense studies, scientists claim,
In chemistries, spectroscopies,

Also physical properties.
Rare earth metals, their salts, too,
Unique and useful things they do.
Lasers, catalysts, and magnets,
Batteries, colors and pigments,
Beautiful for glass, ceramics,
Crafting art and just quaint knickknacks.
Their many useful properties
Lead to some resource scarcities.

China controls the rare earth market,
Their prices claim the business, yet
Other countries have ample ores:
They could scale up domestic stores.
Increased demand will change their stance
Put the worldwide scene in balance,
Supply rare earths from other places.
Set the market on fair basis.
One country should not be the best,
The other countries must contest.
Rare earths we need for many uses,
Monopoly would cause abuses.
No one should monopoly make:
The world's security's at stake.

Heavier Elements:
 The Actinides
 Elements: Actinium, Neptunium,
  Americium, Berkelium, Californium
   Atomic symbols Ac, Np, Am, Bk, Cf
   Atomic numbers 89, 93, 95, 97, 98

The Actinides

A series of 15 metal elements,
Atomic numbers 89-103,
Starting with actinium, symbol Ac,
Then thorium Th, protactinium Pa,
Uranium U, neptunium Np,
Plutonium Pu, americium Am,
Curium Cm, berkelium Bk,
Californium Cf, einsteinium Es,
Fermium Fm, mendelevium Md,
Nobelium No, and finally lawrencium Lr.
All radioactive, similar chemistries
From common 6f orbital electrons.
All rare, do not occur naturally,
Except U, which can be mined as
Ore, refined as yellow-cake.
Plutonium Pu-239 is created from
U-238 for nuclear weapons
And nuclear reactor fuel.

Actinides have very few practical uses,
Mostly in nuclear energy reactors,
Americium in ionization-type
Smoke detectors, thorium in gas mantles.
But no biological roles.
All isotopes radioactive,
Most too dangerous for human exposure.
Stay away from the actinides.

Actinium

Actinium a silvery soft metal,
Highly radioactive, glows in the dark.
Occurs in trace amounts in uranium ore.
Made synthetically in cyclotrons and
From other human-controlled activities.
Atomic number 89, symbol Ac.
Ac named as Greek for beam, ray.
Actinium is start of actinide series of
15 elements ending in 103-lawrencium.
Much like lanthanides, actinide chemistry
Mostly based on +3 oxidation state.
Ac-225 appears in neptunium decay chain,
Good research source of alpha rays, neutrons.
Ac-227 with medical imaging uses.
Otherwise no practical applications
For actinium compounds,
Except in physics labs.

Neptunium

Highly toxic, poisonous,
Pyrophoric, catches fire spontaneously.
Neptunium has little practical use.
Atomic number 93, symbol Np.
Np a heavy metal radioactive,
An actinide, first transuranium element,
Named for Neptune, the planet between
Uranus and Pluto, distanced from our sun.
Hence between uranium, plutonium, in atomic order.
Discovered 1940 bombarding uranium in cyclotron,
Now recovered in nuclear waste.
Np metal with three allotropic forms.
Np chemistry has five oxidation states, +3 to +7.
Np isotopes numerous, most short-lived,
Fissionable but not useful for bombs,
Unlike neighboring elements U and Pu.
Np-239 is intermediate in breeder reactor
Converting U-238 to Pu-239 for A-bombs.
Np-237 most stable, decays in chain
Via francium and actinium.
Np-237 decay chain ends in
Stable thallium-205, unlike most
Other radioactive elements that
Decay to stable lead isotopes.
Which is worse for humans?
Insidiously poisonous thallium, or
Radioactive pyrophoric neptunium?

Americium

Named by Glenn Seaborg in 1944 of
University of California, Berkeley.
He also named elements 96, 97, 98,
Curium, berkelium, californium,
And named the actinide series.
Americium an actinide, an alpha emitter.
Symbol Am, atomic number 95
Americium congruent with lanthanide europium,
Hence both named for continents.
Am very radioactive, major isotope Am-241,
Half-life 432 years, useful alpha source,
Good for ionization smoke detectors.
One microcurie of Am-241 encased in ceramic,
Safe enough to use in homes and offices,
Though a problem in waste disposal.
Commercial amounts of Am produced from
Spent nuclear fuel uranium or plutonium.
Of course Am is also found in
Atomic and hydrogen bomb debris.

Berkelium

Discovered by cyclotron bombarding
Alpha particles at americium-241,
By Glenn Seaborg's group in 1949 at
University of California in Berkeley,
Similar history for californium.
Atomic number 97, symbol Bk.
Silvery metal in actinide series,
Like other actinides with +3 chemistry,
But Bk unusual for odd +4 chemistry
In organometallic complexes.
Bk very rare, no practical uses,
No commercial uses or market,
No biological role except
Extremely dangerous from radioactivity,
Much like californium.
Bk produced in Oak Ridge, Tennessee
Was used to make element 117
In Dubna, Russia in 2010, in a
Joint US-Russia collaboration.
Element 117, the last discovered element
Of the Periodic Table's 118 elements,
Was subsequently named tennessine.

Californium

Similar history as for berkelium,
Discovered 1950 by Seaborg's group at
University of California in Berkeley,
UCB provided names for both
Elements berkelium and californium.
Atomic number 98, symbol Cf.
Cf an actinide like berkelium, but
Unlike Bk some Cf isotopes useful to
Create other elements with
Higher atomic numbers.
Cf sources in portable metal detectors,
Devices for finding metal fatigue flaws,
Even oil and water sources underground.
Cf-252 most useful of 20 known Cf isotopes.
Cf-252 half-life 2.645 years,
Only Cf isotope that emits neutrons,
Acts as trigger for nuclear fission reactions
For nuclear power plants and atom bombs.
Cf-252 synthetic like all Cf isotopes.
Boko Haram did not find a stone
Made of Cf-252 in Nigeria.

More actinides

Of the 15 actinides in the series, five are presented in poems above: 89-actinium, 93-neptunium, 95-americium, 97-berkelium, and 98-californium. Six more actinides are presented below, with elements named for people. These are 96-curium, 99-einsteinium, 100-fermium, 101-mendelevium, 102-nobelium, and 103-lawrencium.

Also see below for poems of three other actinides: 90-thorium, 91-protactinium, and 92-uranium. These actinides are components of the U-238 decay chain. 94-plutonium is presented below with uranium again, as atom bomb elements. All the 15 actinides are heavy metal radioactive elements.

Radioactive Elements Named for People

Of the 118 total of chemical elements known today in the Periodic Table, 14 elements have been named for humans. Most are heavy metal elements, highly radioactive, and synthetic. They do not occur naturally, because they have decayed radioactively since their presence on the primordial Earth. Most were discovered by creating them in high-energy physics research facilities, usually with nuclear reactors or particle accelerators. All 14 follow uranium (atomic number 92) as transuranium elements, and lie in the range of atomic numbers from 96 (curium) to 118 (oganesson).

Naming an element for a person honors that person's contributions to our understanding of nature, regardless of that person's direct connection to the discovery or characterization of the element itself. Nine elements are named for scientists who have done research on radioactive elements (curium, fermium, lawrencium, rutherfordium, seaborgium, bohrium, meitnerium, flerovium, and oganesson). Three are named for other scientists (einsteinium, mendelevium, roentgenium). The remaining two are nobelium and copernicium, named for Alfred Nobel and Nicolaus Copernicus.

Other elements of the Periodic Table have been named with place names (like berkelium, californium), for planets, names of mythological figures (uranium, plutonium), for heavenly objects (helium, selenium), from minerals (beryllium, samarium), traditionally (iron, gold), and a variety of other reasons, some eccentric (yttrium, erbium).

See the list below of all 14 elements named for people. List is in order of atomic number. Following are 14 prose poems, one for each of these 14 people-named elements.

Radioactive Elements of the Periodic Table Named for Human Beings

1. Curium, atomic number 96
   named for Marie Curie (1867-1934) and Pierre Curie (1859-1906).
2. Einsteinium, atomic number 99
   named for Albert Einstein (1875-1955).
3. Fermium, atomic number 100
   named for Enrico Fermi (1901-1954).
4. Mendelevium, atomic number 101
   named for Dimitri Mendeleev (1834-1907).
5. Nobelium, atomic number 102
   named for Alfred Nobel (1833-1896).
6. Lawrencium, atomic number 103
   named for Ernest Lawrence (1901-1958).
7. Rutherfordium, atomic number 104
   named for Ernest Rutherford (1871-1937).
8. Seaborgium, atomic number 106
   named for Glenn Seaborg (1912-1999).
9. Bohrium, atomic number 107
   named for Niels Bohr (1885-1962).
10. Meitnerium, atomic number 109
    named for Lise Meitner (1878-1968).
11. Roentgenium, atomic number 111
    named for Wilhelm Röntgen (1845-1923).
12. Copernicium, atomic number 112
    named for Nicolaus Copernicus (1473-1543).

13. Flerovium, atomic number 114
    named for Georgy Flyorov (1913-1990).
14. Oganesson, atomic number 118
    named for Yuri Oganessian (1933-).

Notes on this list:

1. In only two examples, the scientists for whom the elements were named were still alive at time of naming. All other elements were named posthumously. The two elements are named for Glenn Seaborg, who is now deceased, and Yuri Oganessian, who was alive as of July 2023.

2. The same two examples are elements named for people who participated in their discoveries. They are seaborgium, named for Glenn Seaborg, and oganesson, named for Yuri Oganessian.

3. Only two women are on this list of 14: Marie Curie and Lise Meitner, for curium and meitnerium.

4. Only one person on the list won two Nobel Prizes: Marie Curie, also the only person with two Nobel Prizes in Chemistry on any list.

5. All are Nobel Prize winners, except: Nobel, Copernicus, Mendeleev, Meitner, Flyorov, and Oganessian. The last four were eligible and nominated but did not receive a Nobel Prize.

6. Why not elements named Newtonium, Lavoisierium, Paulingium? Linus Pauling received two Nobel Prizes, one in Chemistry and also the Nobel Peace Prize.

Poems of the Periodic Table
Radioactive Elements Named for People

Curium

Named for Marie Curie, chemist of
Polish descent working in Paris, and for
Her husband Pierre, a physicist.
They were awarded a shared Nobel Prize.
Marie is the only winner of
Two Nobel Prizes in Chemistry,
Only one of two women with an
Element named for her.
Pioneer of radioactive elements,
Famous for discovery of radium,
The element which ultimately
Killed her with multiple skin cancers.
Curium atomic number 96, symbol Cm.
In the actinide series.
Isotopes highly radioactive.
Uses are limited to acting as
Source of alpha radiation in
Analytical spectroscopy.
One of these instruments went to Mars
To analyze the chemical composition
Of surface rocks.

Einsteinium

An actinide element in the series.
Atomic number 99, symbol Es,
Obviously named to honor Albert Einstein.
In one year, 1906, Einstein's *annus mirabilis*,
He made three incredible discoveries,
Each of which worth a Nobel Prize:
(1) mechanism of Brownian motion,
(2) explanation of the photoelectric effect,
(3) special relativity and E=m times c-squared,
The most famous equation in science.
He got the Nobel Prize in Physics in 1922
For the photoelectric effect.
Nazis denied the truth of relativity.
Einstein was anti-Nazi and anti-war.
Peaceful humanitarian, he wrote to FDR
About atomic energy, started Manhattan project.
Element Es discovered in H-bomb explosion
In 1952 US tests at Eniwetok atoll
In the South Pacific Ocean.
No practical uses for Es except for
Laboratory studies, such as
To create the first 17 atoms
Of Mendelevium in 1955.
Es highly radioactive, difficult to handle.
Like all synthetic transuranium elements,
Einsteinium is so hot it glows,
Like the memory of Albert Einstein.

Fermium

Named for Enrico Fermi, of
Nuclear physics fame
Chief scientist at Los Alamos,
Manhattan Project for the A-bomb.
In 1942 he built the famous first
Atomic pile in the squash court
At the University of Chicago.
The first controlled
Nuclear chain reaction.
Fermi could build or repair anything.
Fermium an actinide element,
Atomic number 100, symbol Fm.
Fermium like einsteinium was
First discovered in H-bomb debris
From tests in the Pacific in 1952.
No practical uses except for
Scientific research under
Very careful conditions.
All 20 isotopes known for Fm
Are highly radioactive.
Enrico Fermi refugee from fascist Italy.
Far more useful than the element
Fermium named for him.

Mendelevium

Named for Dimitri Mendeleev,
Chief architect of the Periodic Table.
Atomic number 101, symbol Md.
Highly radioactive, as are
All the transuranium elements and
Actinides with atomic numbers
Greater than 92-uranium,
Formed only in particle accelerators
And nuclear reactors. First made
In 1955 from 99-einsteinium-253,
Still synthesized the same way today.
Seventeen isotopes exist, though
Not much Md is actually produced:
No practical use for the isotopes
Except in scientific research.
On the other hand,
The Periodic Table is still much in use.

Nobelium

Nobelium honors Alfred Nobel,
Dynamite inventor and endower of
Most important scientific prizes,
Also Nobel Literature and Peace Prizes.
Nobel's lasting legacy to promote
Scientific and intellectual endeavors
And honor the peacemakers.
Atomic number 102, symbol No.
One of 14 elements named for persons.
Others 96, 99-104, 106, 107, 109,
111, 112, 114, and 118.
All 14 are transuranium elements.
Atomic number 109 is meitnerium,
Named for Lise Meitner, the only
Woman other than Marie Curie,
Of radium fame. Curium is number 96.
All these and others 93-118 are
Highly radioactive, no longer
Occur naturally on Earth, and
Very dangerous on exposure
To humans and other forms of life.
Yet No is named for Nobel,
Inventor of dynamite.

Lawrencium

Ernest O. Lawrence dominated
Nuclear research at UCBerkeley
For many years, promoted "Big Science,"
Invented the cyclotron particle accelerator
To discover new elements by fusing atoms
With particles like alpha rays and atomic nuclei.
Nobel Prize in Physics in 1958.
Run the particles around in circles,
Faster and faster, bang them into atoms,
Some atoms split, make new isotopes.
Some fuse, make new heavier elements.
Lawrence's name on two U.S. National Laboratories:
Lawrence Berkeley and Lawrence Livermore labs.
Element lawrencium, atomic number 103, symbol Lr
Synthesized in 1961 at Lawrence Berkeley Lab,
Not discovered by Lawrence himself, but
Named for him, confirmed by IUPAC in 1992.
Lawrence designed the separation of
U-235 from U-238 for the Manhattan Project,
Leading to the A-bomb and nuclear reactors.
Lawrencium a synthetic heavy metal,
Highly radioactive, interesting chemistry.
Three double contrasts: U-235 for the bomb
And for clean energy; US and USSR compete for
Priority and naming rights of element Lr; and
Odd Lr chemistry, volatile heavy metal.
A volatile metal? What is that?

Rutherfordium

Ernest Rutherford drew the first
Modern model of the atom.
Later modified by his student Niels Bohr,
Included protons and electrons
But not neutrons, not yet discovered.
Rutherford Nobel Prize winner
Made many discoveries about radiation,
Conceived half-life concept in 1899.
Element rutherfordium named for him.
Atomic number 104, symbol Rf
Highly radioactive, like all transuraniums
In the actinide series, short half-life.
No practical uses.
Rutherford himself inspired research
In radioactivity for more than 50 years.
The element honors his name and work.

Seaborgium

Atomic number 106, symbol Sg
Highly radioactive, does not occur naturally.
Honor for Glenn T. Seaborg, US
Nuclear scientist who created
With cyclotron and linear accelerator
Ten transuranium elements,
Atomic numbers in ranging 94-106,
Examples: curium, for the Curies,
Berkelium, californium for his
Place of discovery at UCBerkeley.
His last, 106, was ultimately
Named for him while he was alive.
Seaborg devised the actinide series,
Atomic numbers 89-103, of
Radioactive elements, in parallel
With the rare earth lanthanides,
Both series displayed as adjacent rows,
Each with 15 elements,
On the current Periodic Table.
Seaborg won the Nobel Prize in 1951.
Seaborgium is the first element
Named for a living person.
Only other is oganesson, named later,
The last element, 118, of the table.
Seaborg also transmuted lead to gold
In 1980, but not nearly enough to pay
Expenses for the research projects.

Bohrium

Discovered after synthesis in 1981, named in 1997
For Niels Bohr, Danish theoretical physicist
Who modified the Rutherford atom
To explain electrons orbiting the nucleus,
Interacting electromagnetically with nuclear protons.
Bohrium atomic number 107, symbol Bh
Extremely radioactive, isotopes very short half-lives,
Bh similar to other transuranium heavy metals.
No practical uses except in nuclear research.
Bohr father of the "Copenhagen interpretation"
Of quantum mechanics first postulated in 1925,
Along with friend Werner Heisenberg of Germany.
Both awarded Nobel Prizes in Physics,
The Heisenberg uncertainty principle.
Heisenberg later directed Nazi atom bomb project
During WWII, visited Bohr in Copenhagen where
Bohr was interned in Nazi-occupied Denmark.
Heisenberg pressured Bohr to work
For the Nazis on atomic research.
Bohr refused and ended their friendship.
No element is named for Heisenberg.
Bohrium honors Niels Bohr.

Roentgenium

Roentgenium, a highly radioactive
Synthetic element, not found in Earth's crust.
Atomic number 111, symbol Rg
Named for Wilhelm Roentgen, discovered xrays,
But not a researcher of radioactive elements.
Rg has strange predicted chemical and
Physical properties, expected to be similar
To copper, silver, and gold.
Mendeleev would have said *eka-gold*,
Based on the column in the Periodic Table.
But so far no data yet to tell, then or now.
Hard to measure, does not exist in nature,
Rg must be created, very short half-life,
No practical uses except in research.
Wilhelm Roentgen the first Nobel Prize winner in
Physics in 1901, the element named after him over
100 years later, in 2004.

Copernicium

Atomic number 112, atomic symbol Cn
Highly radioactive heavy metal, like other
Transuranium elements, numbers 93-118.
Copernicium named for Nicolaus Copernicus,
Renaissance astronomer (1473-1543) revolutionized
Our world view by asserting the sun
At the center of our planetary system,
Earth and the planets revolve around the sun.
"*Eppur si muove,*" said Galileo to the Inquisition.
Copernicium no longer exists naturally on Earth,
Must be synthesized in high-energy
Physics laboratories, though ironically
Formed and present in the sun and stars today.
Galileo and Copernicus also no longer exist on Earth,
But their ideas and insights live on.

Flerovium

Discovered in 1998 at the
Flerov Laboratory of Nuclear Reactions
In Dubna, Russia, element named in 2012.
Laboratory named for Russian
Nuclear scientist Georgy Flyorov, who told
Stalin in 1942 to build an atom bomb for USSR,
Possibly the earliest date of the Cold War.
Flerovium atomic number 114, symbol Fl.
Element 105 also discovered at Flerov Institute,
Named dubnium for Dubna, symbol Db.
Research and discoveries of elements 105, 107, and
112-118 all carried out at Flerov Institute.
Flerovium highly radioactive, half-lives of
Fl isotopes only a few seconds.
They don't stay around long.
About 90 atoms of Fl observed:
Heaviest element ever studied chemically.
Shows properties of both a heavy metal and
Noble gas, may actually be gaseous
At standard temps and pressures.
Can't tell with only 90 atoms observed
For only a few seconds.
Bizarre properties: how can the
Heaviest of metals also be a gas?

Oganesson

Last of them all, at least for now,
Atomic number 118 of the Periodic Table.
Discovered by contrived creation in 2006,
Only five atoms observed (possibly six).
A noble gas, radioactive of course,
Like all elements up from 83 bismuth.
Disappeared from the Earth's creation
By radioactive decay to more stable elements.
Created for discovery in an atomic fusion reactor.
Not much is known about oganesson,
Only five atoms observed (possibly six).
Named for the pioneer of heavy elements,
Yuri Oganessian, a Russian nuclear physicist.
The second element named for a living person.
He is still alive.
The first was Seaborgium, number 106,
Named for Glenn Seaborg, American, of
UCBerkeley, who is now deceased.
Oganesson the heaviest of elements,
Atomic mass number 294, symbol Og,
Very short half-life, highly radioactive.
Og might be a solid, not a gas.
It might be chemically reactive.
Still noble, but not much is known about it.
Only five atoms observed (possibly six).

Elements of the Periodic Table
Radioactive Elements Named for Places

Named for cities, states, countries, and planets,
In order in the table: 84-polonium, 87-francium,
    92-uranium, 93-neptunium, 94-plutonium,
    95-americium, 97-berkelium, 98-californium,
    105-dubnium, 108-hassium, 110-darmstadtium,
    113-nihonium, 115-moscovium, 116-livermorium,
    117-tennessine.
All radioactive, with no stable isotopes.
All metals, with no natural occurrence in ores,
Except uranium pitchblende, occurs worldwide.
All made synthetically except uranium, plutonium.
    Atomic numbers in range 83-118, same range as
    radioactive elements named for people.
Six actinides, range 89-actinium to 103-lawrencium:
    Namely, 92-uranium, 93-neptunium, 94-plutonium,
    95-americium, 97-berkelium, and 98-californium.
No biological activity, all toxic to humans
    from their radioactivity.
No practical uses except americium
    in ionization smoke detectors.
Named for:
1. US cities: berkelium, livermorium
2. US states: californium, tennessine
3. US itself: americium
4. Russian cities: dubnium, moscovium

5. German city: darmstadtium
6. German state: hassium (for Hesse)
7. other countries: Poland: polonium, Japan: nihonium, France: francium
8. planets: uranium, neptunium, plutonium

Contrast the 15 place-named radioactive elements with the 14 radioactive elements named for people.

The U-238 Decay Chain

Uranium-238, the major isotope of uranium, accounts for 99.3% of naturally occurring uranium in ores. U-238 will not sustain a chain reaction. Its minor isotope U-235 can undergo a chain reaction at higher concentration than occurs naturally in uranium ore. Both 92-uranium-235 and 94-plutonium-239 at critical mass can produce chain reactions leading to fission and sudden violent release of energy. The first atom bombs were built using these isotopes. By contrast U-238 at all concentrations is relatively safe to handle.

While reasonably stable, U-238 slowly decays naturally through a radioactive release of energy and transmutation of 92-U atoms to 90-thorium atoms. The decay chain continues stepwise through more intermediate radioactive elements down to stable lead Pb-206. The U-238 radioactive decay explains the gradual slow disappearance of natural uranium, and the gradual slow increase in lead in the Earth's crust. Some other radioactive elements such as neptunium also have extensive decay chains.

The U-238 decay chain comprises nine elements (as isotopes), all radioactive except the last, Pb-206. The seven intermediate elements are isotopes of thorium, protactinium, radium, radon, astatine, polonium, and bismuth. See the illustrations on the inside cover of this book. The following section presents poems about the nine elements of the U-238 radioactive decay chain.

Up and Down the Decay Chain

Start with uranium, atomic number 92.
Isotope U-238 (not the fissionable enriched
High grade U-235, critical mass chain reaction).
You know U-238 is radioactive,
But not fissionable, for the bomb it won't do:
From U-238 the bomb cannot be made.

Alpha decay of U-238 gives 90-thorium-234.
Lose a helium nucleus: two protons, two neutrons.
92-2=90, mass 238-4=234, it's simple to see.
But thorium is radioactive itself, gives new sums,
Two beta decays move up two atomic numbers
(Passing through protactinium, 91-Pa-234).
Directly back to uranium, this time mass 234
To 92-U-234. U-234 gives off another
Alpha particle, goes down to
Thorium again, number 90, mass 230.
Result: 90-Th-234 becomes 90-Th-230.

Now four alpha decays, first new mass 226.
Simple math: 90-2=88, mass 230-4=226.
Thus first to radium, number 88, mass 226,
Then to radon, number 86, mass 222,
Then to polonium, number 84, mass 218,
Then to radioactive lead, number 82, mass 214.

But that's not the end:
Four beta decays and two alphas
Bring us to stable non-radioactive lead,
Atomic number 82, mass 206.
That is the end of this radioactive decay chain
From 92-uranium-238 to stable 82-lead-206.
Eight alpha decays and six beta decays.
Count them yourself.

Intermediate radioactive isotopes of decay chain:
Thorium, radium, radon, protactinium,
Polonium, astatine, and bismuth.
Other decay chains, some may
Start with neptunium or with plutonium,
Have these and other elements as
Intermediates, also including
Actinium, francium, and thallium.

The decay chain moves like clockwork.
Half-lives of some of these intermediate
Isotopes are quite short. Some isotopes
Don't stay around long enough to be dangerous.

Uranium-238, very slow decaying,
Producing little radiation over short times,
Safe enough for adults to handle with care.

Uranium crude ore once included with
Other radioactive substances, also simple
Geiger counter to detect alphas,
In Gilbert box set for children
To play and learn about radioactivity
And atomic energy.
Was it safe?
Probably not, cautious minds prevailed,
The children's set was soon discontinued.

Uranium, Part 1

Present in significant amounts
In Earth's crust in mineable ores,
Atomic number 92, symbol U.
Natural uranium comprises
Several isotopes, mainly U-238
With 0.72% U-235 and some U-234.
Uranium isotopes are all radioactive.
U-238 decay chain with other
Radioactive elements as intermediates

Until reaching stable lead-206.
Extremely long half-lives of U isotopes mean
Actual radioactivity exposure is low.
Uranium is chemically stable,
Means it can be handled safely,
Except for toxicity from inhaling
Dust and gas that are radioactive.
Not enough fissionable U-235
To reach critical mass and explode.
U-238 does not fission.
Risk to humans is low except for
Radon gas in U mines, also mine tailings
Used for building concrete foundations.
Natural U-oxide is excellent glaze
For ceramic tile, pottery, 1930s style.
Bright colors, orange Art Deco dinner plates,
Green floor and wall tiles, other colors.
These uses largely discontinued now.
Depleted U is U-238 with
Less than 0.30% U-235, the
Byproduct of U-235 enrichment.
Depleted U very safe to handle.
Gulf War syndrome is controversial.
U is hard and denser than lead.
Depleted U for tank armor and
Armor-piercing munitions, also
Ballast, weight distribution in planes.
U has many relatively safe uses,
Not just bombs.

Thorium

Atomic number 90, symbol Th
All isotopes radioactive, Th-232 very long-lived,
Called "classically stable" like U-238.
Actinide series, as are two others,
Protactinium and U itself, in U-238 decay chain.
First decay in chain is 92-U-238 to 90-Th-234
By alpha decay. Followed by two beta decays
To produce 92-U-234, and so on.
Decay chain ends in 82-Pb-206, stable.
Among radioactive elements only uranium, thorium
Are long-lived, found in Earth's crust.
Th isotopes occur 3X more than U isotopes.
Another Th isotope Th-232 has its own decay chain
Ending in stable Pb-208.
India has much natural Th, little U, unusual.
India developed nuclear reactors based on Th
To provide fissionable nuclear energy,
Unlike all other countries, which prefer U.

Protactinium

Actinide, atomic number 91, symbol Pa
Discovered in 1913 in uranium ores
As decay product of U-238.
Today extracted from spent nuclear fuel
Named "proto" as decay precursor of actinium.
Chemical reactivity based
Mostly on +5 oxidation state.
Forms oxides, halides with Cl and Br,
Organometallic compounds with tetrahedral symmetry.
Pa element 91 sits between 92-U and 90-Th,
Which are abundant and have practical uses.
No practical uses for Pa because of scarcity
And high toxicity,
Toxic because very radioactive.
Like most actinides, dangerous
For humans and all other life.

Radium

Discovered by the Curies,
Purified as metal by
Madame C in 1911.
Atomic number 88, symbol Ra
Radium does not occur naturally
Except as radioactive decay
Product of uranium and thorium
In ores of those metals.
Ra most radioactive non-synthetic element.
Radium an alkaline earth metal,
Chemistry like calcium and barium,
Divalent Ra++ cations easily
Forming Ra chloride and Ra oxide.
All isotopes of radium are
Highly radioactive, decay to
Radon, also discovered by
Madame C., the only winner of
Two Nobel Prizes in chemistry.
Only use for radium today
Is to collect radioactive radon gas for
Inhalation therapy for lung cancer.
Radium causes radioluminescence
In proximity to zinc sulfide.
Once used to paint toys and
Watch dials, until the painters,
"Radium girls," came down with

Cancer and radiation poisoning
From radium exposure.
Madame Curie died of skin and
Bone cancer from radium.
She was also a "radium girl."
We no longer drink "radium water"
Elixirs to improve our health.

Radon

Radon, atomic number 86, symbol Rn.
Both a noble gas and radioactive.
Described around 1900 by
Ernest Rutherford and Madame Curie,
Best known for discovering radium.
Radon is part of the decay chain
Of radioactive elements, like uranium.
The chain passes down from radium.
All elements in the chain radioactive,
Ending with lead, which is not.
The decay chain responsible for
The gradual disappearance of
Radioactive elements,
Since the time the Earth was formed.
But also responsible for the gradual

Increase in lead on the Earth.
Radon causes much distress downstairs
In houses built in certain places in U.S.
Foundations are radioactive from
Concrete made with sand from mine refuse.
In the West and Southwest uranium mines
Spill their tailings, and sand ends up
In basements as a second use.
Radon gas is heavier than air.
Breathing radon is a cause
Of cancer of the lung, because
Its radioactivity creates DNA lesions
In lung cells, altering cells
From normal to cancerous.
(Like carcinogens in cigarette smoke).
Since radon is a noble gas,
It cannot be removed from air by
Chemical reactions.
Radon noble and heavy radioactive gas:
In this case an unfortunate
Combination for humans.

Astatine

Rarest element in the Earth's crust,
Of all naturally occurring elements.
Atomic number 85, symbol At
Occurs in crust only from radioactive decay
Of thorium and uranium.
Only 30 grams of At exist on Earth.
Never seen by naked eye, because very unstable.
Only one electron needed to complete
The outermost shell, easily gained
Or shared, to give At atom its chemistry.
Called fifth halogen, heavier and
Less reactive than iodine.
Name astatine means "unstable."
At not found in nature because so unstable.
Created in 1940s by alpha bombardment
In UCBerkeley cyclotron, for unambiguous discovery,
Following many failed extraction attempts
From ores, where of course it wasn't present.
Many prior false claims of discovery.

Polonium

Gray metal, highly radioactive
Atomic number 84, symbol Po
Po resembles bismuth and thallium as metals,
But chemically like semi-metals selenium, tellurium.
Po's intense radioactivity causes
Radiolysis of chemical bonds, self-heating.
Po metal is hot!
Discovered 1898 by Marie and Pierre Curie,
Named for her homeland Poland.
Po emits weak alpha rays but nevertheless
Po very intense, very radioactive, very toxic.
Labeled "extremely dangerous to humans."
In 2006 Alexander Litvinenko, Russian defector,
Living in England, met with Putin's agents.
Litvinenko drank tea, got very ill,
Ultimately died in hospital.
British doctors at first unable to diagnose
Polonium poisoning, because weak alpha rays,
Internal inside the body,
Are blocked from external detection.
Colorless, odorless Po salts, very toxic.
Made only in special physical plants, two in Russia.
Putin denied culpability, claimed Western plot.
Putin went on to deal with
Other critics, defectors, dissidents,
Also Crimea and Ukraine,
In his own special ways.

Bismuth

Silvery white metal, with luster like pearl nacre.
Atomic number 83, symbol Bi,
Bi metalnown to the Egyptians for jewelry,
Bi compounds for cosmetics like eye shadow.
Highest atomic mass of stable,
Non-radioactive elements,
Including lead, atomic number 82.
But Bi is very slightly radioactive,
With half-life orders of magnitude
Longer than the age of the universe.
In U-238 decay chain,
Bi almost as stable as Pb lead.
Unusually low melting point,
Good for triggering fire detection systems.
Alloys with tin, lead, iron, cadmium,
Fusible Rose's metal in fire sprinklers.
Woods metal melts in hot water
Like gallium, makes trick spoons.
Also unusual among metals:
Solid Bi floats on melted Bi,
Like ice in water.
Bismuth subsalicylate may soothe
Gastrointestinal and digestive problems,
Products Pepto-Bismol and Kaopectate.
Non-toxic bismuth is better than lead
For fishing weights, birdshot, riot bullets,
Though bullets still heavy enough to kill.

Lead

Symbol Pb for Latin *plumbum*,
Lead the plumber's metal,
Pb very heavy,
Lead weights to plumb the depths.
Lead is soft, ductile, malleable,
Relatively low melting point,
Good for solder and pewter.
Atomic number 82, the highest
Of the non-radioactive metals,
Often the end-product of
Radioactive decay chains
Transmuting radioactive elements.
Lead shielding to block xrays
And other high-energy radiation.
Lead aprons at the dentist and at
Radiation therapy clinics, labs.
Lead bricks at the atomic reactors
And radiation detecting devices.
The two sides of lead and radiation.
Many other uses: batteries, gasoline, bullets.
Lead compounds in paints and pigments,
Sometimes in children's hard candy,
Softening agent in telephone cords.
Don't chew on them!
Pb unfortunately highly toxic,
Causes nerve damage.
The two sides of lead in common use.

Atom Bomb Elements: uranium, plutonium
Atomic numbers 92, 94

Uranium, Part 2

U-235 when enriched to 30% or so
Can be used in atomic reactors.
Enriched to 90% for bombs.
Uranium U-235 very fissionable,
Chain reaction at critical mass
For A-bombs, nuclear power plants,
Trigger for thermonuclear H-bombs.
U-235 is less than 1% of natural U,
Major U isotope is U-238, not fissionable.
U must be enriched in U-235 to sustain fission.
Oak Ridge diffused U as
Uranium hexafluoride UF6
Through membranes. U-235-F6 diffuses
Only very slightly faster than U-238-F6,
Requires repeated steps, long times, to
Enrich natural U in U-235.
Nowadays high-speed centrifuges,
Faster but still slow and expensive.
Major project feasible only for
Governments dedicated to
Building the Bomb.
A matter of national prestige,
Also nuclear aggression and defense.

Plutonium

Radioactive metal, many isotopes
And allotropes, hard to handle,
All forms highly radioactive
And pyrophoric, burst into flame
Spontaneously in air.
Atomic number 94, symbol Pu.
Nevertheless Pu finds application
In nuclear weapons and
Nuclear energy power plants.
Breeder reactors with U-235 and U-238
Convert U-238 to Pu-239, using
One neutron capture and one beta decay
To make 93-neptunium-239,
Then one more beta emission to convert
93-Np-239 to 94-Pu-239.
Pu-239 as fissionable as U-235
For the same energy purposes.
Pu-239 first produced in large amounts
At the Hanford, WA plant,
Manhattan project and for
Fat Man atomic bombs
Detonated in New Mexico desert and
Dropped on Nagasaki in August 1945.
A-bomb WWII in the Pacific and start of
The Cold War and nuclear arms race.

Atomic Bombs

Fat Man and Little Boy
The first atomic weapons.
Fat Man, plutonium-239 from Hanford,
Tested at Trinity site, New Mexico,
Close along the Jornada del Muerto,
The Trail of the Dead Man,
On July 16, 1945. It worked.
Little Boy, uranium-235 from Oak Ridge,
Dropped on Hiroshima
On August 6, 1945. It worked.
A second Fat Man dropped on Nagasaki
On August 9, 1945. It also worked.
They worked as designed, ended the war,
But also failed as practical weapons,
Morally, politically, even militarily
Unacceptable.
More A-bombs exploded in tests,
Over 120 A-bombs made,
And thousands of H-bombs.
Fallout from testing nuclear weapons
Poisoned people, animals, fish,
Plants, made territory uninhabitable.
Atmospheric and underground testing,
Radioactive fallout, the Cold War,
The arms race, stockpiles of nuclear
Weapons, terrorist threats.
The legacy of atomic bombs.

Fission and Fusion

When an atom splits in two, it's fission.
Two smaller elements are made from one.
Splitting for some elements is mission,
But others may require a cyclotron.

Much energy released when atom splits.
Some elements require a stimulus.
A neutron gives U-235 the fits:
Its nucleus explodes spontaneous.

U-235 is special: when it goes,
Its nucleus releases three neutrons.
And that's the start of all our hopes and woes.
Chain reactions make power and A-bombs.

The other parts when U-235 splits:
Krypton-92, barium-141.
Other atoms will also fall to bits
When hit by neutrons, Pu-239 is one.

Fusion means two nuclei collide
And fuse to make a new heavier atom.
Great energy is needed to provide
Their protons and neutrons to fuse as one.

Three facts about this fusion to relate:
1. In universe all elements were fused.
First hydrogen, then helium create,
In stars and supernova still enthused.

2. Heaviest elements of the table
Long ago decayed, now fused in labs
In reactors, accelerators able,
Use energy, make only tiny dabs.

3. The H-bomb is our worst delusion.
Exploding U or Pu provides the heat
To start the earthly process of H-fusion.
Mimic the sun, perhaps to change our fate.

Cold fusion is a dream yet to fulfill.
Fusion needs much hot energy to prime.
The idea: get back more energy still,
But how to start without condensed sunshine?

Fission and fusion are two sides to tell
The story of atomic energy.
Most atoms do not either fizz or gel.
They just behave with well-known chemistry.

Section 2: People

Bad Luck Blues

[refrain from "Born under a Bad Sign", recorded by Albert King in 1967]

My gal's so mean, she don't ever call.
My gal's so mean, she won't take my call.
If it wasn't for bad luck,
I'd have no luck at all.

When we go out, she wants to have a ball.
When she goes out, she gotta have a ball.
If I didn't have bad luck,
I'd have no luck at all.

I'm so down, I'm heading for a fall.
She put me down, so I'm heading for a fall.
If it wasn't for bad luck,
I'd have no luck at all.

I'm a little short, and my gal's real tall.
I'm too short, and my gal's way too tall.
If I didn't have bad luck,
I'd have no luck at all.

I want some loving, but all she do is stall.
I gotta get some loving, and all she want is stall.
If it wasn't for bad luck,
I'd have no luck at all.

I beg for some loving, and she just makes me crawl.
She's so mean to me, she always makes me crawl.
If I didn't have bad luck,
I'd have no luck at all.

I wanna go to the bar, but she goes to the mall.
I'd take her out, but she goes shopping at the mall.
If it wasn't for bad luck,
I'd have no luck at all.

I went to the bar anyway, just got into a brawl.
She wouldn't patch me up, told me not to bawl.
If I didn't have bad luck,
I'd have no luck at all.

My gal's got my money, she made my pockets small.
She got all my money, she made my pockets small.
If it wasn't for bad luck,
I'd have no luck at all.

Second

It's enough for a lifetime.
Always being second.
Born a second child of my parents.
Almost flunked second grade.
Primary school not for me,
Secondary school more appropriate.
My monopoly card: Won second prize
In a beauty contest.
Came in second in voting for
Junior high class president.
Community foot race, second place,
Only a second later than first.
Sang only harmony in Glee Club,
They voted me vice-president.
Played second base in baseball,
But only the second string.
Graduated not valedictorian
But number two in class,
Can you say salutatorian?
Runner-up in science fair, got one-year
Scholarship to second-rate school.
My name in second column
Of company letterhead,
Second name from the top.
Published professional articles,
Always listed as second author.
Bet on the Derby. Horse came in second,

Paid $2.60 on $2 bet. Winner paid $7.40.
Smoke cigars, good brands,
But cheap factory seconds.
Second alternate juror at major trial.
Spouse says, "I'm almost ready to go."
"Hold on a second," I say.
"I'm here ready to be held."
Spouse says I was spouse's second choice.
Spouse was my second choice two.
We live many seconds north latitude,
And many seconds west longitude.
Drive a second-hand car.
Boss in England seconded me
From executive branch to
Inferior secondary department.
Rode second-class on European trains.
In office the boss said, "Wait for a second."
I said "No need to wait. I'm here now."

Great Men of Cigars

Winston Churchill set the tone:
He led Great Britain with his will.
He smoked cigars long as a bone.
Winston's type we now call Churchill.

Sigmund Freud smoked 'til he died.
Found penis symbols near and far.
Jung noted Freud's cigar. Freud sighed:
"Sometimes a cigar is just a cigar."

Words Rudyard Kipling would summon
When writing a poem, a joke:
"A woman is only a woman,
But a good cigar is a smoke."

Bill Clinton and Monica also enjoyed
Cigars: He smoked long ones, not blunt.
Bill liked the good taste it employed
When he stuck one in Monica's …. front.

Pres. JFK smoked cigars, and took
Pre-war Havanas from Castro.
It is not known whether he stuck
One into Marilyn Monroe.

George Burns, when very old, still joked.
When asked if he would date a lot,
Said "No," waving the big cigar he smoked.
"This is the stiffest thing I've got."

"You bet your life" said Groucho Marx,
He puffed, blew smoke, acted a liar.
But he was careful with the sparks
To not set his eyebrows on fire.

Whenever Boston Celtics won,
Red Auerbach lit up a cuban.
He puffed, waved it, 'til it was done.
The other team had no true fan.

Mike Jordan used a cigar cutter,
Then could not play: he cut his thumb.
Later he was heard to mutter,
"I lost millions! Boy, was that dumb!"

Mark Twain once said in conversation:
"They say my smoking is a crime.
But I indulge in moderation,
Just smoke one cigar at a time."

These are some well-known men who smoke.
There are still more, like Schwarzenegger.
You could be one also - no joke!
Whether you're poet or a beggar.

Double Oboe Concerto

The oboe is a very difficult instrument to play.
Musicians say it makes you crazy to play it.
Other musicians say you must have been crazy
To start with,
Otherwise you wouldn't have chosen the oboe.
The double reed is hard to blow,
Just to get a sound.
The air pressure builds up in your head
And probably causes brain damage.
The oboist must carve his own slim fragile reeds.
The double reed is difficult to carve correctly,
With many frustrating failures.

I play the oboe.
I am an oboist.
I have mastered the fingering, the many
Alternative fingering variations.
I carve my reeds quickly and skillfully.
My reeds perform perfectly.
I am a master oboist,
Perhaps the best in the world.
I learned the circular breathing techniques
Of the great oboist Heinz Holliger.
I am not crazy yet.
I am in demand by orchestras around the world
To play the oboe repertoire.

I solo, play the great concertos of
Bach, Vivaldi, Albinoni.
I have made many recordings,
To critical acclaim.
Including the great double oboe
Concerto of Tomaso Albinoni,
On recordings and in live concerts.
Unfortunately it requires a second oboist,
And I have no peers among oboists.
I must play concerts with inferior second oboists.
I have recorded the great double oboe concerto,
Playing both parts in tandem.
The magic of the recording studio.
But I am determined to play it in concert,
Live, with an orchestra, as a sole oboist,
Alone.

At first I alternated playing both parts,
Switching between the parts quickly.
It is possible to do that
In places where the parts are not simultaneous.
But in many places the oboe parts overlap.
Then I tried to finger quickly enough
To play both parts simultaneously,
Back and forth, shortening the notes,
To trick the ear of the listener
Into thinking there were two of us.
This sometimes works well enough, except
It compromises the music as written,

Shortening some overlapping notes
In both parts, especially in the third movement
Which has long passages where
Both oboes play simultaneously
With very fast fingering.
I astonished listeners with my virtuosity,
But I was not satisfied.

I got a second high-quality oboe.
I held it alongside the first.
I trained very hard, until
I could blow both at the same time
And make good sounds.
Now I had to finger both at the same time
To produce a different melody
Or harmony on each oboe
Simultaneous with the other oboe.
It's hard to blow one oboe, doubly hard with two.
Am I crazy yet?
Since the oboe, like other woodwinds,
Needs both hands to finger
All the notes on one oboe,
How could I play
Two oboes at the same time?

I devised small felt pads, with rubber backing,
That I pasted onto the underside
Of my fingers, near the palm.
Now with the second oboe to the right

Of the first, I could finger both oboes
At the same time,
By stretching my fingers, sometimes unnaturally.
I could finger the left oboe (first oboe)
With my right-hand fingertips, and
Finger the second oboe on the right side
With the felt pads at the base of the fingers.
Likewise for my left hand fingering
The first oboe with the left-hand felt pads
And the second oboe with the left fingertips.
I was very proud of this invention.

I practiced stretch-fingering and blowing hard.
I blew hard and practiced more fingering.
Hours and hours a day,
Stretching my fingers and blowing hard
For many months,
Until I was ready to perform the
Great Albinoni double oboe concerto
In front of an audience
With orchestra
As a soloist.

But I did not perform.
In dress rehearsal my fingers cramped
On the stretched passages, paralyzed.
My fingers were permanently frozen.
I blew so hard I suffered a stroke.
I became crazed, irrational.

They said I was crazy.

Reluctantly I agreed.
The great Concerto for Two Oboes
In F major, opus 9
Of Tomaso Albinoni
Must be played in concert
By two oboists.

How Many Was That Again?

The worst time of my life was the
Year they were all teenagers
At the same time.
When we two got married,
We said we wanted twelve children.
We had the first, and
After a few months,
Said maybe ten would be enough.
After the second child, we said
Perhaps just seven or eight total.
After the third, we said
Just five or six.
After the fourth, we said
"Let's stop here."
The six of us, two adults and

Four wonderful kids,
Were reasonably happy together,
In spite of scars, poison oak,
A few broken limbs.
Traveling, a year living in Europe,
Kids full-time in local schools.
All children learned the language.
Months in Australia, the
Older two traveling,
The younger two in boys' school.
Music, sports, camping
Worked well for a time.
Traveling we had to count
To six, to check that
We were all present together.
We often said:
How Many Was That Again?
Now the children are gone,
All four university graduates
With advanced degrees,
Gone to their other lives and spouses.
We're now counting grandchildren.
How Many Was That Again?

Aaron and Ara

They were born in the same minute
In the same hospital
In adjacent delivery rooms.
Witnessed one of them being born,
Heard the cries from the next room,
At the same time as ours.
Saw the nurse take our baby to the center room
For cleaning and weighing
And Apgar evaluation, no doubt.
Stayed to comfort the mother of our child,
Then followed the nurse after a few moments.

Nurse at the scales and examination table,
Another nurse next to her
With another baby.
Nurses amused by two baby boys,
Looking alike, born in same moments.
Nurses with their backs turned to me,
Both busy with newborns.
Nurses similar from the back:
Short dark hair, body shape, wearing scrubs.
Which nurse was which?
Which baby was which?
Which baby was ours?
Nurses helping each other, holding babies,
Putting name bands on babies' wrists.
Nurses left, each carrying a baby boy

Back to his mother.
My very careful intense scrutiny
From across the room,
Virtually certain which one was ours.

Later Aaron and Ara grew up
In the same neighborhood,
Went to the same schools,
Looked alike, acted alike,
Both blond, upright, gentle dispositions.
Coincidentally similar names.
People couldn't tell them apart.
Some thought they were twins,
Very similar countenances, looks, behavior.
They became friends in kindergarten
And primary school,
Both very intelligent, mild-mannered, polite.
Visited each other's homes.
Hard to say who was who,
Even around the dinner table.

Later Aaron's hair got darker,
Like our other children's.
Ara stayed blond, like his siblings.
They drifted apart, as children do.
Still later, Aaron went to Yale,
Where, tall, quiet, dark-haired,
He was often confused with his older brother,
Also at Yale, in grad school,

Four years different in age.

Ara disappeared from our lives.
Aaron is still around.

Freshmen Heavyweight Crew

I am a modest person.
I don't accept credit for what I didn't do.
Sometimes I don't accept credit
Even if I did a creditable thing.
At school one day I received a letter
Asking me to come by the
College sports club room
To receive my jacket patch and certificate
For participating on the championship
Freshmen heavyweight crew.
At Ivy League schools crew is
Very important.
I had watched crew races before.
I did not otherwise participate in crew.
I had never been in a crew scull
Or touched a scull oar.
I had no desire or aptitude for crew.

I did not crew.
I was thin and not muscular,
Not very athletic.
Definitely not a heavyweight.
I was a second-year graduate student,
At least six years older than a freshman.
Definitely not a freshman.
I went to the college sports club room.
It was mid-day, only one person present.
He was enormous, possibly an
Offensive tackle in football
Or a heavyweight wrestler.
He looked at me curiously.
I told him why I was there and
Showed him my award letter.
He noted my name, looked into a large
Box of envelopes, found one
With my name, and checked to see that
It contained a jacket patch and certificate.
He stood up, looked me
Straight in the eye, and said
"Congratulations."
He shook my hand vigorously.
It hurt for two days.
Later I went to the school's registry office.
I looked up students with my name.
Surprisingly there were three of us.
I called to tell them what happened,
To give the envelope to the correct person.

The first guy was a foreign student
Who didn't know what crew was.
The second disdained all college sports.
I agreed with both of them.
I could not give the award envelope to anyone.
I never heard from the sports people again.
I still have the award, more than 50 years later.
I have never displayed the envelope's contents.
I am a modest person.

80 and More

Eyesight almost gone
But hearing not too bad
Except for tinnitus
Teeth a few lost
Others going
What happened to taste?
At least there's still chile.
Aches and pains
Knees, hips, elbows
Arthritis in joints and hands
Muscle pain, cramps at night
Drugs keep cholesterol in check
Also blood pressure pills
Take blood thinners too
Dizzy when standing, walking
Look out for a fall
Can't get off the floor?
I don't like this at all.

Co ... py

"Was it clear?" she asked,
Over the phone.
"Yes, mostly," I said.
"Just very slightly cloudy,
And slightly greenish."
"Probably greenish from bile," I said, "and
Cloudy from epithelial cells sloughing off."
She was silent. Patients aren't
Supposed to know about bile,
Let alone epithelial cells.
Was I a patient? I wasn't ill.
More like a client or customer.
Later, the procedure.
"Was it painful?" a friend asked me afterwards.
"No, not even uncomfortable.
I was unconscious."
Or rather, awake and talking throughout,
They told me, but no memory of it.
Apparently that's the effect of fentanyl.
The two days before were very unpleasant, however,
If not a little disgusting.
"Well, what was the result?"
"I'm more-or-less okay, don't need
Another one for five years."
"That's not going to happen,"
I said fifteen years ago.

"I'd rather die."
(No I wouldn't.)
And so far I haven't.

Lost Boys

Our neighbors had two children,
Boys, about two years apart.
Both very intelligent, clever,
Near geniuses.
Early computer hobbyists
In the earliest days of home computers.
With doting parents, both professionals,
A supportive home life.
They did brilliantly in school.
The older boy outgoing, well-spoken.
The younger more reserved, quieter.
The older boy went to engineering university,
Then into the tech industry.
A successful professional life.
The younger boy disappeared.

The older boy lost his brother.
He argued with his parents, lost contact,
Moved far away with his wife and two children.

His wife refused to meet them,
Refused to let the two young children
See or talk with the grandparents.
The older boy lost his parents and his brother.
Never heard from his brother again.
He became a lost boy.

The younger boy was not seen again,
Nor heard from, except for a
Very short local phone call
To his mother, message unclear.
The police could not find him.
Local authorities could not help.
Private detectives, at great expense,
Turned up nothing.
He disappeared for thirty years.
Then a police detective in New York City
Called to tell the parents
Their lost boy had been found.
Exhumed from a grave in potters' field,
Dead at least thirty years,
Died before the age of 20.
Dead all these years,
He was a lost boy.

The parents had two children,
Brilliant, intelligent, healthy,
Wonderful childhoods,
But they became two lost boys.

Section 3: Miscellany, Fun and Serious

The Purpose of Life:
Science and the Chicken

The child says, "The chicken is a bird."
The parent answers, "The chicken is not a bird.
It is an odd form of life, a monster.
At best it is an honorary bird."
The child says, "But it has feathers and wings."
The parent replies, "Birds fly. Did you ever see
A chicken fly? Chickens cannot fly."
The child says, "The news announcer said that
Petaluma, a town in northern California, has a
Chicken-flying contest every year."
(True: The chicken is forced to walk
Out on a raised diving board, jump off,
And flutter safely to the ground.
The winner is the chicken that lands
Farthest from the front of the board.)

The child says, "My teacher asked us,
'Which came first, the chicken or the egg?'"
"That's easy," replies the parent:
"Obviously the answer is the egg, which was laid
By a bird similar to but not the same as a chicken.
The egg would be a mutant, different from the
Mother bird. The egg hatched into a chicken,

An animal not like the mother bird,
Though similar in some ways.
This may happen many times,
Making many small changes.
The chicken is a new form of life,
Made for human use as food,
Not a real bird that could live in the wild.
The chicken does not exist in the wild.
Did you ever hear of a wild chicken?"
The child says, "The news said a truck crashed
On the freeway near Los Angeles.
It was carrying crates of live chickens.
Some crates split open, and chickens escaped.
They have been living under the freeway overpasses
And breeding for several generations now.
They are feral. Doesn't feral mean wild?"
The parent says, "Until the coyotes get
The last one."

The child asks, "What is the purpose of
The chicken? Is it just for food?
For people and coyotes to eat?"
The parent answers: "The biologist says
That the chicken is the egg's way
Of making another egg."
The child asks, "What does that mean?"
The parent responds, "In order for any animal, bird,
human, or any other form of life to sustain itself,
There must be a future.

The future requires having offspring.
The body of the animal is a means
Of creating offspring.
The biologist would say,
'The soma lives and survives
To make gametes to allow the gametes
To survive and reproduce,
To create the next generation, where
The genes of the gametes produce the
Soma, to enable passing on the genes
Into the future.' The body of the animal is
The vessel and everything else needed
To reproduce the gametes."
Survival and reproduction.
This is one way to understand
The purpose of life and the chicken.
The Purpose of Life.

Bird Song

They sing in the garden at daylight,
Starting at dawn with a few notes,
Full-throated in the sunny afternoon.
Chirps, peeps, whistles,
Some quiet as whispers,
Some loud as a scream.

Like speech: mostly vowels,
Especially "long e" and
Short "o" and "i",
Palatals and linguals.
Some consonants for sure: plosives,
Fricatives, but labials few or none.
Often a trilled "r".

Bird lovers know each species has its own song.
Ornithologists recognize local dialects
Between populations of the same species
In different locations.
Occasionally a dislocated bird
Will learn a new dialect from its new neighbors.

Why do they sing?
Maybe to attract mates.
Maybe to defend territory.
Maybe to tell others of a rich source of food.
Like morning insects
Or ripe berries in the afternoon.
Perhaps just to announce themselves
To friends and relatives.

In the morning a robin with a worm,
Later, rock doves terrorizing sparrows,
A murder of crows in the nearby cornfield,
Threatening everyone.
As darkness falls the soft "hwoo" of the owl.

On the Air

Go outside, check the antennas.
Wire antennas don't have much wind load,
Better for our QTH
With occasional 80 mph winds.
The antennas look okay, still standing.
Go back to the shack,
Sit down at the bench.
Turn on the power strip,
Then the power supply.
Turn on the transceiver.
Select a frequency band.
Select the mode: cw for now.
Put on the headphones, and
Connect the straight key.
Turn the knob to tune the vfo,
And listen.
Make sure of the correct
Range of frequencies.
Tune and listen again.
Find a likely clear spot.
Push the antenna tuning button.
Position the key within reach.
Listen again. Hear other cw
Stations chirping in QSOs.
Continue tuning the vfo.
The band seems clear,

With little QRM or QRN.
No fading, no QSB on signals.
Tap the key once, hear a beep,
The transmitter and keyer are working.
Tap out CQ: dahdidahdit dahdahdidah.
Repeat several times.
Tap out dahdidit dit didahdah
didididahdah dahdahdit didahdah dididah.
Repeat twice. End with dahdidah.
Listen. Adjust vfo slightly.
Listen again. If no response,
Repeat the keying again.
Listen carefully, pencil ready.
On the air.

Four Rhymes

Four words in English claimed not to have rhymes: orange, silver, purple, and month

1A. orange
From the Manual of Home Repair
To fix the door that's painted orange,
You'll need a lock, a latch, and door hinge.

1B. orange

> Candy Lover

You like the candy bar called "Orange Chew"?
Then you're a candy-lover, aren't you?

2. silver

> The Shiny Rocket

The old park's rocket ride is silvery still.
But if you ride it, you'll get ill, very ill.

3A. purple

> Pancakes (a la Roger Miller)

Pancakes with grape jelly: very purple,
Taste much better served with maple surple.

3B. purple

> Crayons

My box of crayons has many purples.
I don't need so many. I've got surplus.

4. month

> Elizabeth

Elithabeth, who hath a listhp, can teach
At Thaint Maryth's Thcool for Speeth.
And with her friends goes out to lunth
In a bunth, at least onth a month.

Exercise in Style
(after Raymond Queneau, 1947)

Bus
Noon
Bus
Man
S-bus platform
Younger man
Neck
Hat
Ribbon, not a ribbon, plaited
Feet
Jostle of passengers
Foot stomp
Angry words
Move to seat
Later
Gare
Younger man, friend,
Overcoat, button
Advice
(End)

Reimagining the Alphabet

A is for aardvark
That double aa is compelling.
B is for beehives
Do bees attack aardvarks? Do aardvarks eat bees?
C is for calabash
What the heck is a calabash?
D is for d..d..dead
Happy as the day the dog died.
E is for emergencies
Heart attack, or aardvark on the highway.
F is for fake olds
The opposite of fake news.
G is for global warming
Too hot for aardvarks and people.
H is for heart attack, of course.
Stress, fatty foods, and no aardvarks.
I is for idolatry
Do not worship aardvarks, nor with them.
J is for jail
Avoid jail, infested with aardvarks.
K is for kick
Don't kick the aardvark.
L is for light bulb
Better to see the aardvarks all around us.
M is for Mama
Mama doesn't care about aardvarks.

N is for nose
Nosy aardvark also has a long tongue.
O is for old man
See the old man getting older.
P is for people
Old men, old women are still people.
Q is for quince
What the heck is a quince?
R is for rumor
Someone saw an aardvark today? That's a rumor.
S is for snack
The aardvark is not a tasty snack.
T is for termite
Termites eat wood, aardvarks eat termites.
U is for Uncle Aaron, definitely not an aardvark
Though that double Aar is compelling.
V is for victim
Aardvark on the highway is a victim.
W is for watch
Watch out for aardvarks on the highway.
X is for X-ray
See the termites inside the aardvark.
Y is for yawn
Yawn – no aardvarks at the zoo.
Z is for zee, not zed
The British zoo zed-bra is dead.

Nantucket Limericks

A housewife who lived in Nantucket
Raised a chicken in order to pluck it.
But the chicken ran free
And begged to disagree.
So she picked up a hatchet and struck it.

A young woman who came from Nantucket
Lost her dress when a fierce windstorm took it.
She ran around scanty
In just her red panty.
A tattoo on her breast read "Can Suck It."

The beer stock was low in Nantucket,
So we went up to Boston and drunk it.
We wanted still more,
So we bought out the store,
And we rented a big van to truck it.

Said a sullen young man from Nantucket,
"If there's work to be done, I will duck it."
He was grumpy and slow.
The new boss told him so:
"Go get a big lemon and suck it."

An old gal who lived in Nantucket
Would relieve herself into a bucket.
One night in distress
She made such a mess,
She said to herself, "Ah, go shuck it."

There's a lusty young man in Nantucket,
His organ so long that he stuck it
Folded into his shorts,
But his girlfriends, good sports,
Are happy to help him untuck it.

A loose woman plies trade in Nantucket,
Wears a dress with an open front placket.
On busy nights she would test,
When she took off her dress,
If pressed to the wall she could stuck it.

We bought a fat hen in Nantucket.
It was live; with a sharp axe, we struck it.
In the oven it stank,
The burnt smell was so rank.
We forgot that we needed to pluck it.

A dim-witted young man from Nantucket
Found a junky old radio and plugged it
In the wall of the shower stall,
And in no time at all
He lit up brightly while kicking the bucket.

A young girl and young boy in Nantucket
Played hand ball quite well, and each struck it.
But the girl won the bet,
And the boy could not get
His hands on her, said "Ah, the heck with it.

More Limericks

Carol's Party

Carol's Party on such a nice day,
Outside in the garden we stay.
The hot people inside
Crowded close side by side.
But they nevertheless had to pay.

Hot Honey
(Or, a lady gets a jar of chile-flavored honey)

For a gift, maybe rather got money,
Though your husband may think it's not funny,
But perhaps it is true
When you're both pitching woo,
That he calls you his "Sweet" and "Hot Honey."

Norton

A man in Virginia named Horton
Made wine from the native grape Norton.
The red wine was so fine
It won tastings done blind.
He called it, of course, "Horton Norton."

Torque Board

There's a man who came here from New York
Had a board made of wood and not cork.
When he spun it around
It flop-flipped upside down,
And he just could not get any torque.

Chile Hat

A man who drank grapefruits as juices
Knew that it soothed hunger and bruises.
"If it wouldn't look silly
On my head I'd wear chile,
Just to prove that I've been to Las Cruces."

Nirvana

Old friends sit around drinking beer,
Smoke cigars, talk old cars, have no fear.
We're seeking Nirvana,
(It's not Texarkana.)
Too bad we can't get there from here.

Xmas Gift Poems

Short poems originally written to attach to Xmas presents. On Xmas morning each person in turn opens a gift, first reading aloud the attached poem. Some are simple rhymes, some humorous, some serious, a few limericks and riddles. All are short. The following are selected from more than 200 poems written over the last few years.

    Gift: socks in cigar box
Hold a cigar between your toes.
Take a puff: it's a yoga pose.
Cigars are useful, so I'm told.
When your toes get very cold.

    Gift: kit with anti-virus mask
What is it?
It's in a kit.
It may be knit.
What is it?
Will it itch?
Don't ask.
Depends on stitch.
To wit: what is it?
Not hard to ask.
It's a _____.

Gift: hot peppers
Your mother grew these in the yard.
It really wasn't very hard.
She brought them just for you to taste.
Eat them carefully, not in haste.

Gift: jar of peppers
No surprise here –
You get them every year.
The only question that you've got
Is whether they're sweet, mild, or hot.

Gift: Spanish dictionary
Another language you both know.
Together check your diction.
When you see how words will go,
 Digame: Tell me fact from fiction.

Gift: three pairs of socks
Three and two equal six.
And you've got ten
To cover with two,
Or is it one, a choice of three?

Gift: bag to cook potatoes
Potatoes are not a good Xmas present.
Potatoes do not grow in a bag.
It is better to store potatoes in air,
Not in a bag.
Then what the heck is a potato bag?

Gift: stuffed doll, a wood sprite
A hobbit is a bad habit.
You put an elf on the shelf.
An orc is a dork.
A wizard is not a lizard.
You leave a dwarf at the wharf.
What do you do with a wood sprite?
Oh, you'll find out, all right.

Gift: celery crisper
You get no salary,
Though hardly a calorie,
When you accelerate
To the celery crate.

Gift: potholder
You think it's hot but it's not.
It's light and tight, all righter.
Everyday it's good to use,
And daily food it won't abuse.
Do not look for it above.
As a cook, look near the stove.

Gift: short colorful dress
It is patchwork, multicolored, bright,
But maybe too small to fit.
Try it. It may be all right.
But where could you possibly wear it?

Gift: small travel sewing kit
This Xmas gift is neither good nor bad.
It's just so-so.

Gift: Tony Hillerman novel
Navajos and New Mexico desert:
Some things you didn't know you need.
The pull of the story will keep you alert,
Because it's a very good read.

Gift: Joy of Cooking book
Joy! Joy! Joy!
Cook! Cook! Cook!
Boy oh boy!
It's a good book.

Gift: red chili powder
Put it in your pants
To keep away bees and ants.
Don't put it on your head.
"That would be silly," they said.
Put it in your pocket.
Try it before you knock it.
Don't put it in your socks.
It'll feel like walking on rocks.
Put it in your food.
Now you've got it! Tastes very good.

Gift: bottle of porter flavored with coffee
When you want a drink, my dear,
A cup of coffee or a beer,
No matter what the weather,
How about the two together?

Gift: game of Sorry
A middle-age love affair
Of long-standing years,
Means never having to say you're sorry,
And you can win without tears.

Gift: single-player game
This game will not win you fame or wealth.
It is best to play it after dinner.
You have to play it by yourself.
The question is: Will you be a loser,
    Or a winner?

Gift: privately printed poetry book
You will certainly have to accept it,
Because you can't politely reject it.
And please don't refuse it,
Even if you don't use it.
It's a work of rational sanity,
Or perhaps just natural vanity.
In it you may occasionally delve,
And then to the bookcase to shelve.
So far there are only four, not twelve.